Full **STEAM** Ahead!

Technology Time

Energy Everywhere

Cynthia O'Brien

CRABTREE
PUBLISHING COMPANY
WWW.CRABTREEBOOKS.COM

Title-Specific Learning Objectives:

Readers will:

- Identify types and sources of energy.
- Understand the difference between renewable and nonrenewable sources of energy.
- Identify the main ideas in the book—that technology helps us collect energy, and that renewable energy is better for the environment—and their supporting details.

High-frequency words (grade one)	Academic vocabulary
can, come, from, into, make, that, use	electricity, energy, fuel, renewable, source, technology

Before, During, and After Reading Prompts:

Activate Prior Knowledge and Make Predictions:

Have children stand up. Instruct them to act out what it looks like to have a lot of energy. Now ask them what it looks like to have no energy. Give them a few seconds to demonstrate each idea.

Ask children to explain what energy is. Accept all answers and then summarize them by defining energy as the power that makes things move, work, and grow. Then ask children what they think they'll be learning about in a book called *Energy Everywhere*.

During Reading:

After reading pages 8 and 9, ask children to explain what a renewable energy source is. Have them give examples of sources that are renewable, and those that are not. Which do they think would be better for the environment?

After Reading:

Have children create exit cards to capture what they learned. Hand out index cards or small pieces of paper. Tell children to write the answers to these three questions, which you can write on the board:

- What is one new thing you learned?
- What fact did you think was most interesting?
- How can you use what you learned in your own life?

Author: Cynthia O'Brien

Series Development: Reagan Miller

Editors: Bonnie Dobkin, Janine Deschenes

Proofreader: Melissa Boyce

STEAM Notes for Educators: Bonnie Dobkin

Guided Reading Leveling: Publishing Solutions Group

Cover, Interior Design, and Prepress: Samara Parent

Photo research: Samara Parent

Production coordinator: Katherine Berti

Photographs:
Alamy: Avalon/Construction Photography: p. 21
Shutterstock: Legenda: p. 8; N. Antoine: p. 10; Juan Enrique del Barrio: p. 20

All other photographs by Shutterstock

Library and Archives Canada Cataloguing in Publication

Title: Energy everywhere / Cynthia O'Brien.
Names: O'Brien, Cynthia (Cynthia J.), author.
Description: Series statement: Full STEAM ahead! | Includes index.
Identifiers: Canadiana (print) 20190231572 |
  Canadiana (ebook) 20190231580 |
  ISBN 9780778772644 (softcover) |
  ISBN 9780778771791 (hardcover) |
  ISBN 9781427124579 (HTML)
Subjects: LCSH: Power resources—Juvenile literature. |
  LCSH: Force and energy—Juvenile literature.
Classification: LCC TJ163.23 .O27 2020 | DDC j333.79—dc23

Library of Congress Cataloging-in-Publication Data

Names: O'Brien, Cynthia (Cynthia J.) author.
Title: Energy everywhere / Cynthia O'Brien.
Description: Ontario ; New York : Crabtree Publishing Company, [2020]
  | Series: Full STEAM ahead! | Includes index.
Identifiers: LCCN 2019052909 (print) | LCCN 2019052910 (ebook) |
  ISBN 9780778771791 (hardcover) |
  ISBN 9780778772644 (paperback) |
  ISBN 9781427124579 (ebook)
Subjects: LCSH: Power resources--Juvenile literature.
Classification: LCC TJ163.23 .O27 2020 (print) | LCC TJ163.23 (ebook) |
  DDC 621.042--dc23
LC record available at https://lccn.loc.gov/2019052909
LC ebook record available at https://lccn.loc.gov/2019052910

Printed in the U.S.A./032020/CG20200127

Table of Contents

Crabtree Publishing Company
www.crabtreebooks.com   1-800-387-7650
Copyright © **2020 CRABTREE PUBLISHING COMPANY**. All rights reserved. No part of this publication may be reproduced, stored in a retrieval system or be transmitted in any form or by any means, electronic, mechanical, photocopying, recording, or otherwise, without the prior written permission of Crabtree Publishing Company. In Canada: We acknowledge the financial support of the Government of Canada through the Book Publishing Industry Development Program (BPIDP) for our publishing activities.

Published in Canada
Crabtree Publishing
616 Welland Ave.
St. Catharines, Ontario
L2M 5V6

Published in the United States
Crabtree Publishing
PMB 59051
350 Fifth Avenue, 59th Floor
New York, New York 10118

Published in the United Kingdom
Crabtree Publishing
Maritime House
Basin Road North, Hove
BN41 1WR

Published in Australia
Crabtree Publishing
Unit 3 – 5 Currumbin Court
Capalaba
QLD 4157

Energy All Around

You cannot see it, but **energy** is everywhere. We need energy for everything we do.

A toaster uses energy to provide the heat needed to make toast.

Energy makes this lamp glow in the dark.

Do you use a computer for school or to watch videos? It needs energy to turn on.

What is Energy?

Energy is the power needed to make things work, move, or grow. A light bulb needs power. So do you! Your body gets energy from the food you eat.

A kite uses energy from the wind to fly high into the sky.

This blender uses **electric** energy to make a smoothie.

We use heat energy to cook our food.

Kinds of Energy

Energy comes from many different **sources**. Some sources are **renewable**. They never get used up. Other types of sources cannot be used more than once. They are not renewable.

Some energy comes from **fuels** such as coal, gas, and oil. These fuels power our cars, trains, and airplanes. Once these fuels are used, though, they are gone forever.

Energy can also come from wind, water, and the Sun. This type of energy is renewable. Renewable energy is better for Earth.

Energy and Technology

Technology is anything people create to make life easier, safer, and more fun. Some new technologies help us to use energy that is better for Earth.

Most trucks, cars, and trains use oil and gas to make them go. These fuels **pollute** the air.

New technologies can make cleaner fuel from plants such as rapeseed (above). These fuels can be used in new types of cars and engines.

Using Sun Energy

Technology can help us use the Sun's energy. This energy can heat a house and make lights glow.

These big **panels** collect the Sun's energy and turn it into electricity.

Lights like these collect energy from the Sun during the day. The energy turns them on at night.

This **battery** collects energy from the Sun. It then provides energy to the phone.

Using Wind Energy

Windmills are a technology used to collect energy from the wind. The energy is turned into electricity.

Old windmills like these were once used on farms.

A pinwheel is like a small windmill. It turns when you blow on it.

The large windmills used today are called wind turbines. They can be hundreds of feet tall.

wind turbine

To make a lot of energy, people build wind farms with lots of wind turbines.

15

Using Energy from Earth

The inside of Earth is very hot. The heat makes hot water and steam. Both are sources of energy.

Sometimes hot water and steam burst from inside Earth.

Buildings called **power plants** use Earth's hot water to create electricity.

Pipes can bring up heat from Earth to keep plants in a greenhouse warm.

Using Water Energy

Water is another source of energy. Power plants turn falling or flowing water into electricity.

Water flowing in a river has a lot of energy.

Big dams like this collect water and then send it pouring out. The falling water is turned into electricity.

Waterfalls can also be used to make electricity.

Good for the Planet

Renewable sources of energy are best for the planet. They do not pollute the environment. They never run out.

People are working hard to make technologies more Earth-friendly. This charging station has panels that collect energy from the Sun. It uses this energy to provide electricity to electric cars.

Technologies such as wind turbines, batteries, and power plants help us use renewable energy. It is up to us to choose to use energy that comes from renewable sources.

wind turbine

The owners of these houses use small wind turbines to get electricity from the wind's energy.

Words to Know

battery [BAT-er-ee] noun An object that supplies electricity

electric [ee-LEK-trik] adjective A type of power or energy

energy [EN-er-jee] noun The power that makes things move, work, and grow

fuels [fewls] noun Materials that are burned for heat or power

panels [PAN-ls] noun Thin, flat pieces

pollute [poh-LOOT] verb To make dirty or unsafe

power plants [POW-er plants] noun Buildings with machines inside that generate power

renewable [ri-NOO-abuhl] adjective Able to be used again

sources [sors-es] noun Where something comes from

A noun is a person, place, or thing.

A verb is an action word that tells you what someone or something does.

An adjective is a word that tells you what something is like.

Index

About the Author

Cynthia O'Brien has written many books for young readers. It is fun to help make a technology like a book! Books can be full of stories. They also teach you about the world around you, including other technologies—like robots.

To explore and learn more, enter the code at the Crabtree Plus website below.

www.crabtreeplus.com/fullsteamahead

Your code is:
fsa20

STEAM Notes for Educators

Full STEAM Ahead is a literacy series that helps readers build vocabulary, fluency, and comprehension while learning about big ideas in STEAM subjects. *Energy Everywhere* helps readers identify the main idea of the book and give examples of energy types and the technologies that harness them. The STEAM activity below helps readers extend the ideas in the book to build their skills in technology, science, and language arts.

Capturing the Sun

Children will be able to:
- List ways we get energy from the Sun.
- Explain how technology helps us capture the Sun's energy.
- Build a simple solar oven by following directions.

Materials
- Building a Solar Oven Direction Sheet
- Paper plates and snack supplies (e.g., chips and cheese for nachos, or fixings for s'mores)
- Materials for solar oven: shoe box, aluminum foil, black paper or thin poster board, plastic wrap, tape, glue stick, ruler, sticks to prop open the box

Guiding Prompts
After reading *Energy Everywhere*, ask children:
- What sources does energy come from?
- What are some ways we use energy?
- How does technology help us use energy?

Activity Prompts
Have children turn to pages 12 and 13 and review the ways that power from the Sun is used. Tell children that another name for Sun power is "solar power." Remind children that technology helps us capture the Sun's energy.

Tell them that they will have a chance to make their own solar technology: a solar oven. They'll also get to cook a snack!

Hand out the direction sheet. Remind children that when they follow directions, they should read all of the steps first to see if they understand them. Give them time to read the directions and ask questions.

Divide children into groups of 3 or 4 and hand out supplies. Give them time to build the oven. Model each step at the front of the class. When they are done, take the ovens to a sunny spot. Give each team supplies for the snack that will be cooked. Note: Food may take 15 to 30 minutes to cook.

Extensions
- Ask children if they can figure out reasons for the materials used. Prompting questions include: Why did we use foil? (Reflects the heat of the Sun.) What does black paper do? (It absorbs heat.) Why did we put on plastic wrap? (It keeps the heat inside the box.)

To view and download the worksheets, visit **www.crabtreebooks.com/resources/printables** or **www.crabtreeplus.com/fullsteamahead** and enter the code **fsa20**.